MW00964613

SCIENCE WORLD

LIVING THINGS

KATHRYN WHYMAN

Stargazer Books

New edition published in the United States in 2005 by:
Stargazer Books
c/o The Creative Company
123 South Broad Street
P.O. Box 227
Mankato, Minnesota 56002

Editor: Katie Harker

Designer:
Pete Bennett – PBD

Picture Researcher:
Brian Hunter Smart

Illustrator: Louise Nevett

Printed in UAE

Library of Congress Cataloging-in-Publication Data

Whyman, Kathryn.
Living things / by Kathryn Whyman.--New ed.
p. cm. -- (Science world)
ISBN 1-932799-27-3 (alk. paper)
1. Biology--Juvenile literature.
I. Title. II. Science world (North Mankato, Minn.)

QH309.2.W48 2004
570—dc22

2004041824

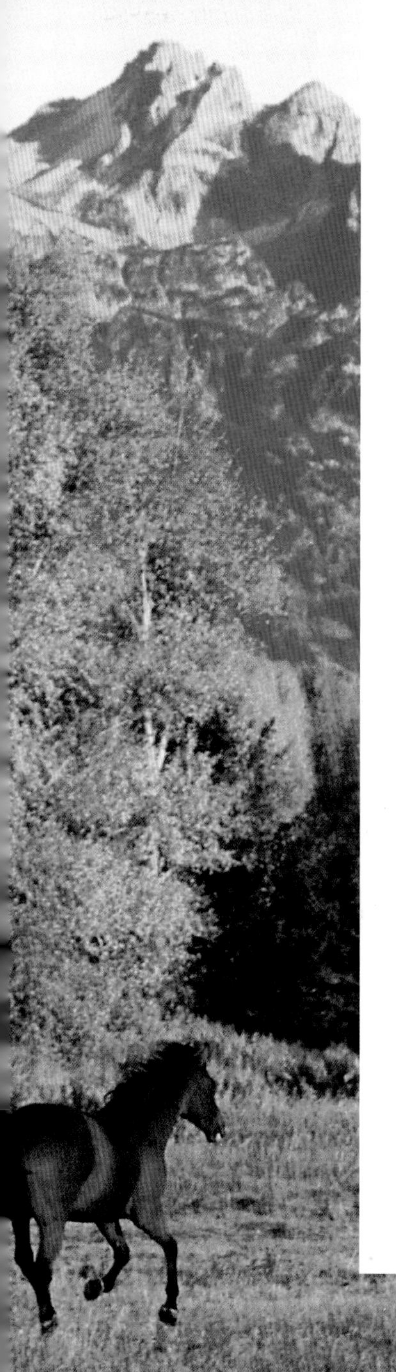

Contents

INTRODUCTION

Our world is home to many millions of living things, from the hot, dry deserts to the frozen polar regions, at the depths of the oceans, and even inside the bodies of other creatures! All living things are specially adapted to enable them to survive in their different environments.

Living things may look very different, but they all share some basic characteristics: they need oxygen and "nutrients" for nourishment; they move and grow; they get rid of waste substances; they react to things around them; and they can also reproduce.

Polar bears are specially adapted to live in the cold extremes of the polar regions.

In this book you will find out how living things cope with the changes that are constantly taking place around them. You will learn how living things have grown and evolved over millions of years. You will also discover that some living things communicate, and that all living things depend on each other for their survival.

Right: Chameleons change the color and pattern of their skin, to blend in with their surroundings and protect themselves from attack.

Lions are well camouflaged in the hot, dry plains of Africa.

Is it alive?

Imagine you were looking at a wax model of a child. The model might confuse you at first. It may look very realistic and even share many of the physical features of a child. But you would soon be able to tell that the waxwork was not a living thing.

A real child can move. A child needs to eat and breathe to survive, and excretes waste materials, like feces and urine, from its body. A child can see and hear things happening nearby and will act accordingly. Eventually, the child will grow and may even have children of its own. The wax model may appear to be very lifelike, but it can do none of these things.

Living things can move about, sometimes very quickly, like these human sprinters.

Living things need nutrients for nourishment. This hummingbird is feeding on flower nectar, a good source of sugar. In turn, the flower has been feeding on sunlight and nutrients from the soil.

We plant many kinds of trees, shrubs, and flowers in public places and in our gardens.

THE VARIETY OF LIFE

There are about two million different types of living things today! Although they share similar features, there are also great differences between them. People have found it useful to sort living things into groups. The two largest groups are the plant and the animal kingdoms.

Within these groups there are many different types, or "species." So scientists divide the plant and animal kingdoms into smaller groups. Animals are first divided into those that have backbones, the "vertebrates," and those that do not, the "invertebrates." Plants have also been divided into many different groups that distinguish between their structure and their leaf or flowering cycles.

The animal kingdom

Most animals are invertebrates (without a backbone). For example, earthworms, butterflies, spiders, and crabs are just a few invertebrates. Vertebrates can be put into five groups: fish, amphibians (vertebrates that spend part of their lives in water and part on land), birds, reptiles, and mammals. Mammals are the only animals that produce milk to feed their young.

Giraffes, the tallest living animal, are found in the open grasslands of Africa.

Many different types of plant species can be found in this meadow.

EVOLUTION

Where did all the different types of living things come from? Nobody knows for certain, but many scientists think that plants and animals have gradually developed, or "evolved," over millions of years. As they have changed, they have become better adapted to survive.

For example, millions of years ago plants had no flowers. They relied solely on the wind and the rain to transport their pollen to other plants, so that new seeds could develop. But over time, plants developed simple flowers to attract insects. Insects accidently pick up sticky pollen as they feed on flower nectar, and carry it to nearby plants, helping the flowers to reproduce.

Human beings may have evolved from apelike animals. These developed the ability to stand upright on their feet, so that they could then use their hands for other things. Over millions of years they learned how to use tools, and then became successful hunters.

Fossils are the remains of plants and animals that have been preserved in rock. They give us clues about the life of ancient living things. Scientists can work out the age of fossils by dating the rocks in which they are found.

Insect-pollinated plants usually have fragrant, brightly colored flowers to attract insects.

CELLS

All living things are made up of tiny building blocks called "cells." Cells are too small to see without a microscope. Some living things have one cell, but the human body consists of about a hundred trillion cells!

Almost all cells contain a nucleus. The nucleus is very important as it controls everything that happens inside the cell. Around the nucleus is a jellylike substance called "cytoplasm." Here, lots of chemicals are stored. Around the cytoplasm is a very thin "skin" called the cell membrane. This holds the contents of the cell together and controls what enters and leaves the cell. Plants and animals have different types of cells. These cells are all designed for a particular job.

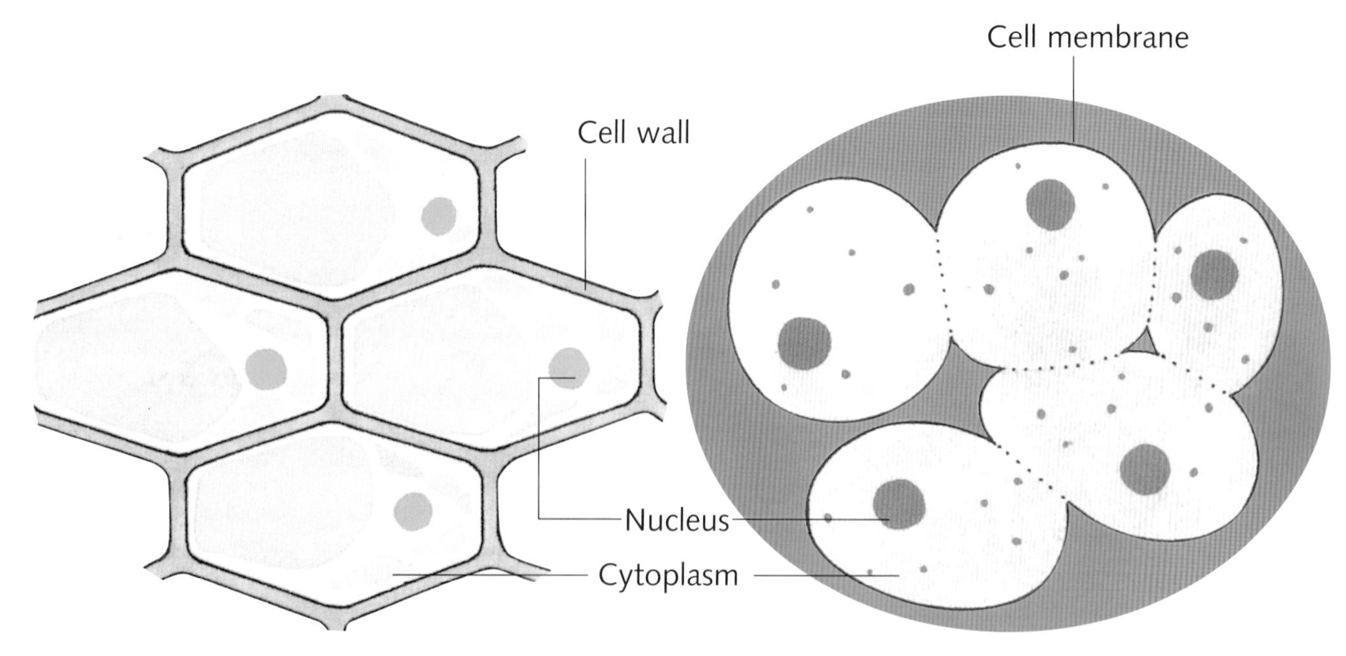

Plant cells
Plant cells are each surrounded by a "cell wall" made of cellulose, which gives the plant structure and support. They each have a nucleus and cytoplasm. Plant cells usually have a very regular shape.

Animal cells
Animal cells don't have any cell walls (just a cell membrane). This is because animals use other ways of supporting themselves, such as skeletons. Animal cells are usually irregular in shape.

Special cells cause pigmentation in these zebras' coats, creating a striped camouflage.

A leaf under a microscope clearly shows the structure of a plant cell.

FEEDING AND BREATHING

Living things need certain substances to move, grow, and keep themselves alive. Animals eat plants or other animals as their main source of nutrition.

This food is chemically complicated and must be reduced to simpler materials by a process called "digestion." Chemicals in the body break down the nutrients in food, which the body then uses for growth and energy. Waste materials are then excreted by the body.

Photosynthesis
Plants and animals depend on each other for feeding and breathing. Plants take carbon dioxide from the air and absorb water and minerals from the soil. They use the sun's energy to convert these simple substances into sugars and starches and produce oxygen. This is called "photosynthesis."

Animals, like gorillas, use plants as a source of food. They also breathe in the oxygen that plants release. Animals produce carbon dioxide which they breathe out, and they excrete waste water and chemicals from their bodies. Plants depend on all these substances.

Plants need oxygen in order to carry out certain life processes. Plants use photosynthesis to produce oxygen but they also absorb oxygen from the soil (through their roots) and from the air (through small holes in their leaves).

At night, plants take oxygen from the air because there is not enough sunlight for photosynthesis to take place. During the day plants produce their own oxygen—much more than they need. This excess oxygen is released into the air.

Plants provide oxygen for humans and animals to breathe.

Squirrels need to eat about a pound of food a week to maintain an active life.

GETTING FROM PLACE TO PLACE

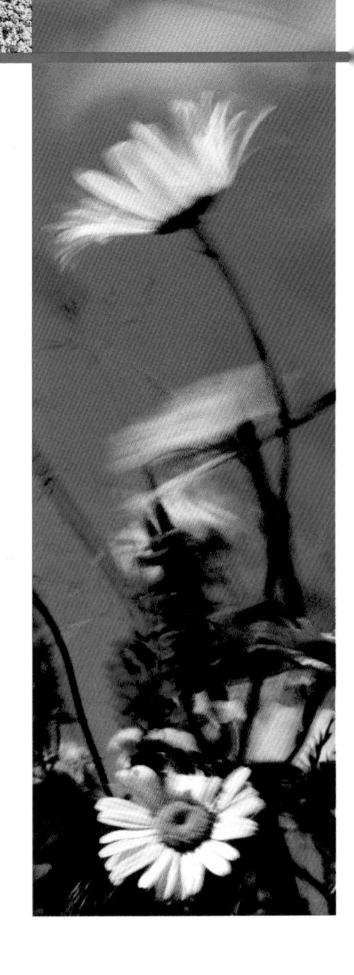

Most animals have to move to find their food and to avoid predators and other dangers. Many of these animals have muscles to help them move. Muscles help fish to swim, birds and insects to fly, and many animals to walk and run.

Plants move by growing in different directions. When water is in short supply, plant roots grow deeper into the soil to find it. Shoots grow taller to find more sunlight. Plants also need to move their pollen and seeds. Pollen and seeds are "dispersed"—spread around—from their parent plant so that they have their own space to grow. They may be carried by animals, water, or the wind.

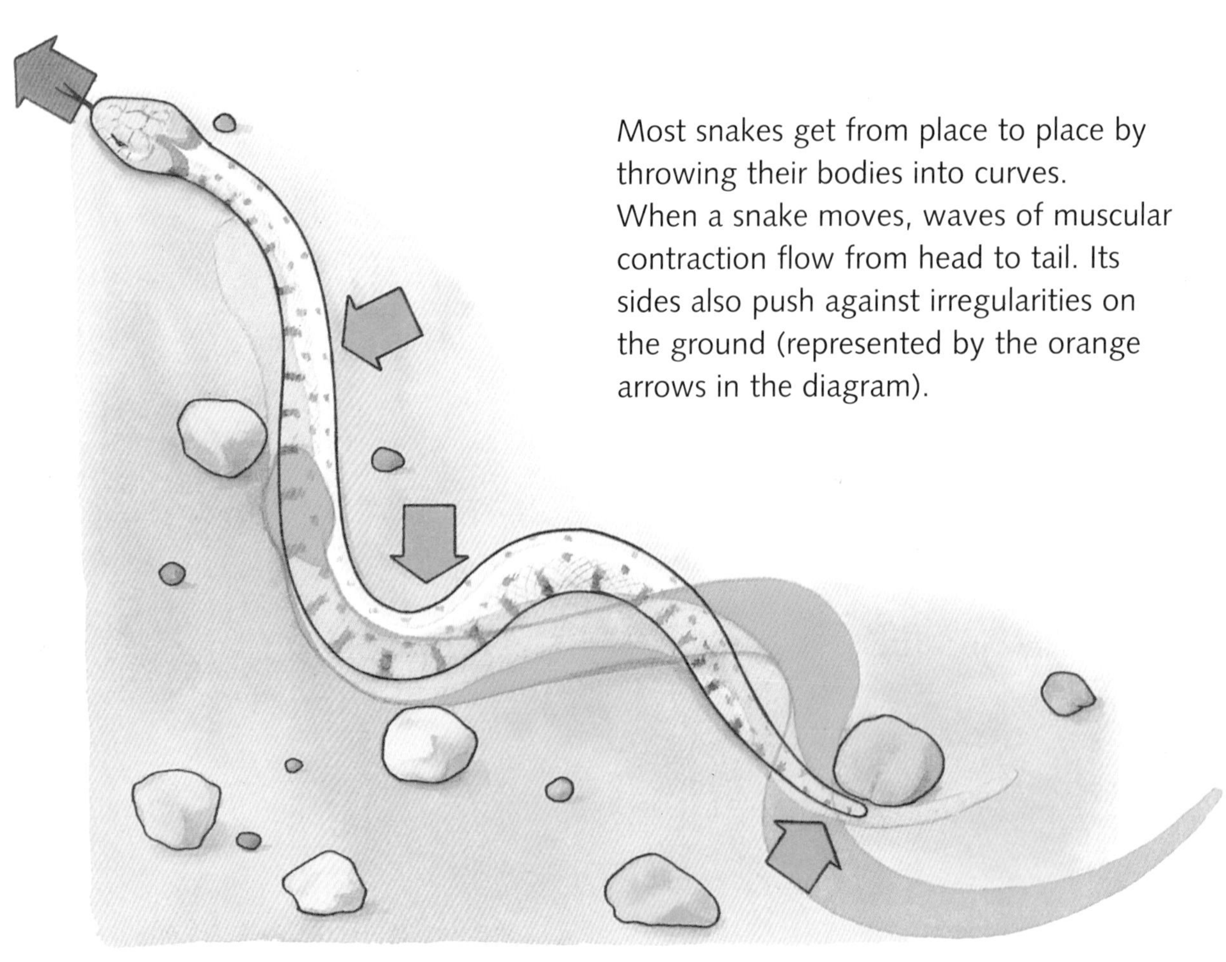

Most snakes get from place to place by throwing their bodies into curves. When a snake moves, waves of muscular contraction flow from head to tail. Its sides also push against irregularities on the ground (represented by the orange arrows in the diagram).

Virtually every part of a cheetah's body has been adapted to maximize its running speed.

Climbing plants, like ivy, use walls for support, and grow toward the sunlight.

A CHANGING ENVIRONMENT

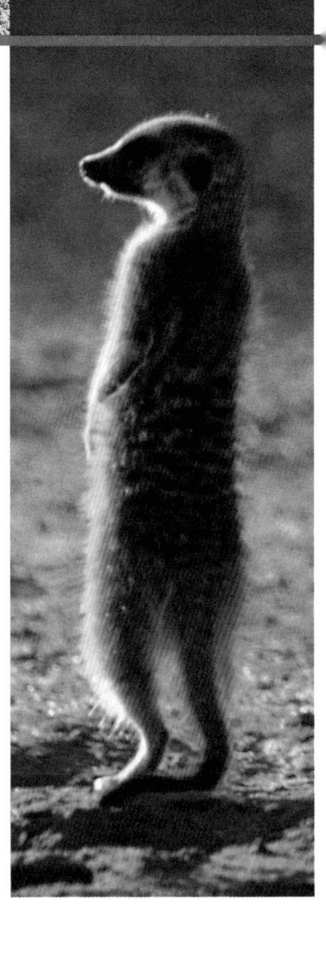

In order to survive, living things must be able to detect changes in their environment and react to them. This includes adapting to changing weather conditions and the threat of living things that may approach them.

Animals have up to five senses—sight, smell, hearing, taste, and touch—to help them to detect changes around them. A dog may react to the smell of food by running toward it; an earthworm will react to light by burrowing underground; and a chameleon changes its skin color and patterns to match its background. Plants don't have senses of this kind, but they can still detect changes in their environment.

This sea anemone reacts to touch...

... by retracting its tentacles.

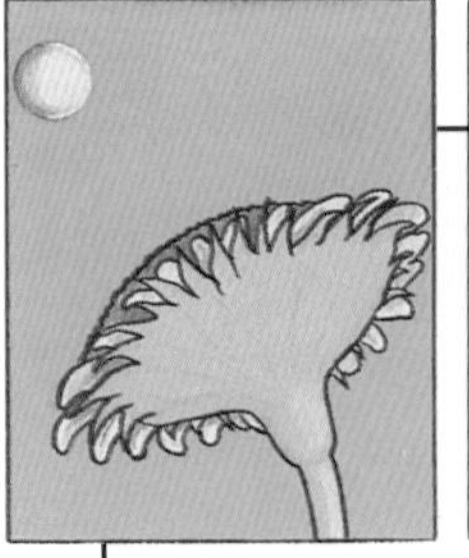

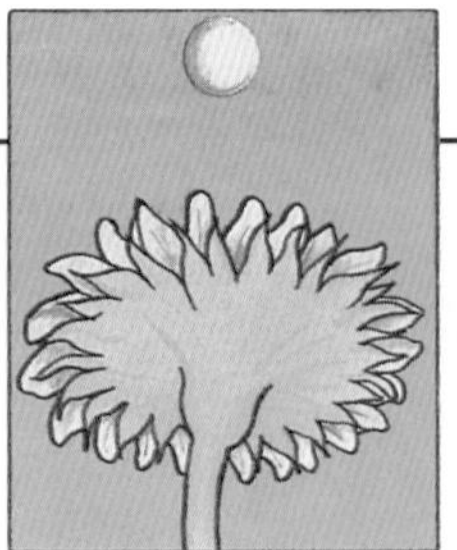

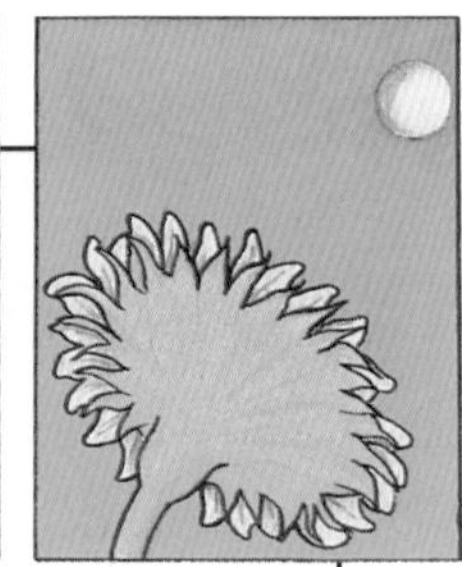

Plants can react to certain changes in their environment. We have seen how plants grow toward light and water. This sunflower gradually turns as it follows the path of the sun during the day. Some flowers, such as lilies, open during the day and close at night. Plants are remarkably sensitive to heat, light, drought, certain chemicals, and gravity, but they have no special organs for detecting change.

An Arctic fox's fur turns from brown to white in winter to camouflage it in the snow.

REACTING TO CHANGE

When we react to a changing environment, conditions *inside* our bodies change. Your body's main instinct is survival, so it reacts to potentially dangerous situations.

When you exercise, most of the energy in your body is released as heat energy, and your temperature rises. To lower your body temperature, blood moves to the surface of your skin where it can cool down. Water also evaporates from your skin as sweat, helping to cool your body. During the winter months, physical changes can also help an animal to hibernate, while food is scarce.

In the fall, many birds migrate to a warmer winter climate.

In the winter months, some animals, like dormice, are no longer able to find food. Their bodies are able to adapt, and they "hibernate." The animal "sleeps" for the winter. Its body gradually gets colder, its heartbeat slows down, and it breathes less often. In this condition, animals use little energy. They can survive without eating and live off stores of fat inside their body.

It is important to drink after exercise to replace fluid lost through perspiration.

GROWTH AND REPRODUCTION

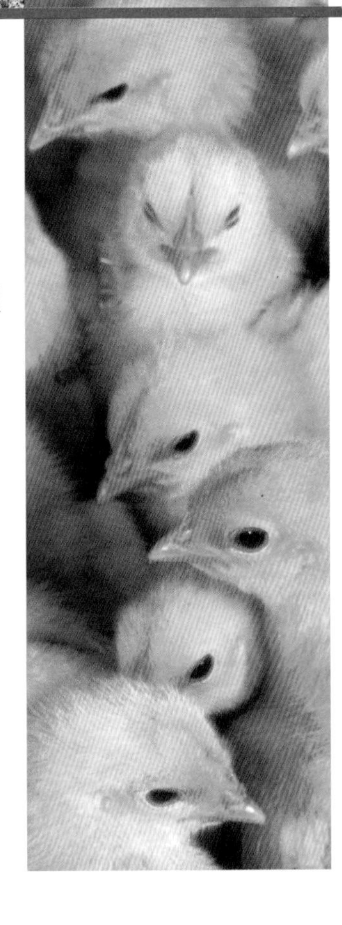

As plants and animals develop, they grow and get larger and heavier. How does this growth take place? We know that living things take substances, like food, into their bodies. Some of these substances become part of body cells. Cells get bigger until they cannot grow any more and they divide into two. As more cells are formed, a living thing grows.

When living things are fully developed they are able to "reproduce" and create new members of their species. Most living things in the animal world reproduce when special "sex cells" from the mother and father join together. As this cell grows and divides, a new living thing develops.

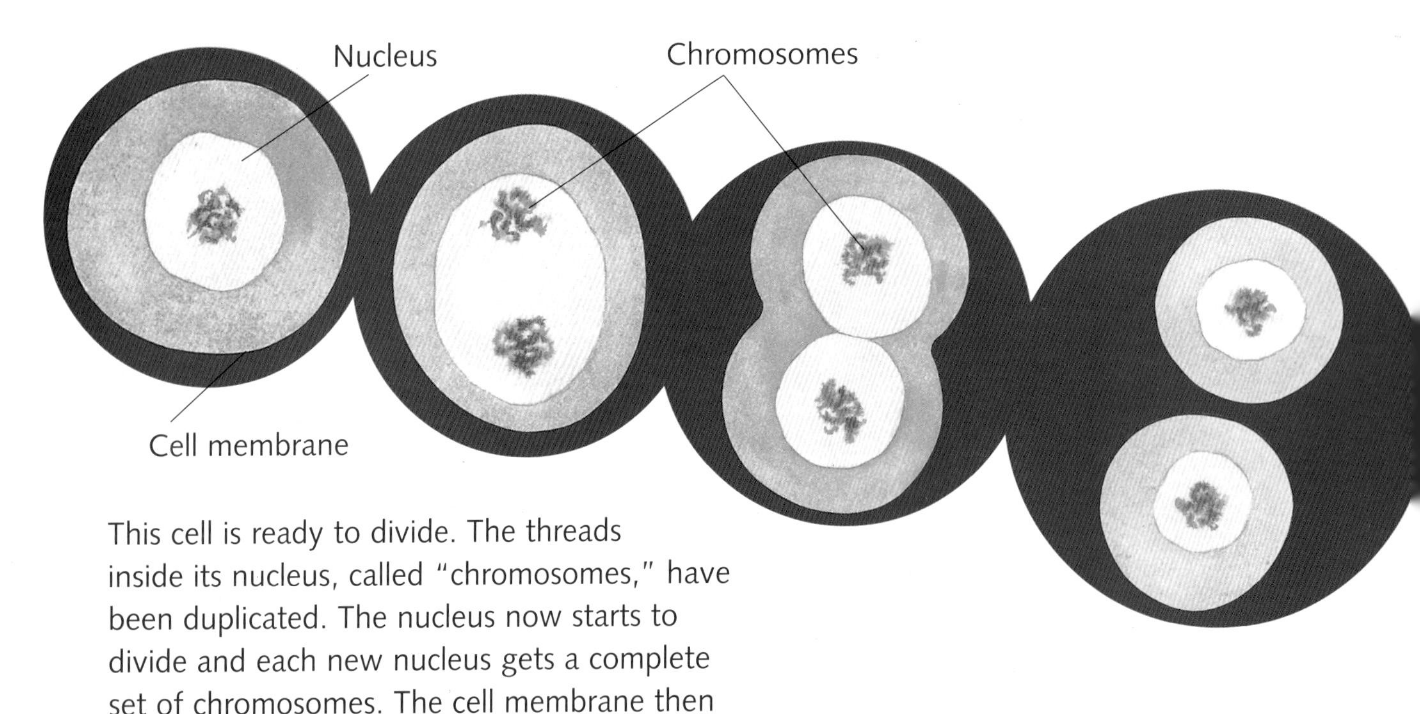

This cell is ready to divide. The threads inside its nucleus, called "chromosomes," have been duplicated. The nucleus now starts to divide and each new nucleus gets a complete set of chromosomes. The cell membrane then divides to form two separate cells. These are identical to the original cell with the same number of chromosomes. They will grow and eventually divide themselves into two new cells.

This calf will grow and develop for up to seven years before it becomes fully grown.

Yew trees take 1,000 years to mature, while pine trees only take 30 years.

COMMUNICATING

Many living things can pass information to each other—they "communicate." The expression on the face of a cat may show that it is angry or frightened. Bees perform complicated dances to tell each other where to find good sources of food, and monkeys show each other affection by grooming each other and cuddling their young.

Some animals communicate by producing chemicals. Cheetahs mark out their territories by spraying urine and scent onto plants. Birds and mammals also communicate by making sounds. Humans, as the most intelligent of all animals, have developed the most complicated and successful form of communication—speech.

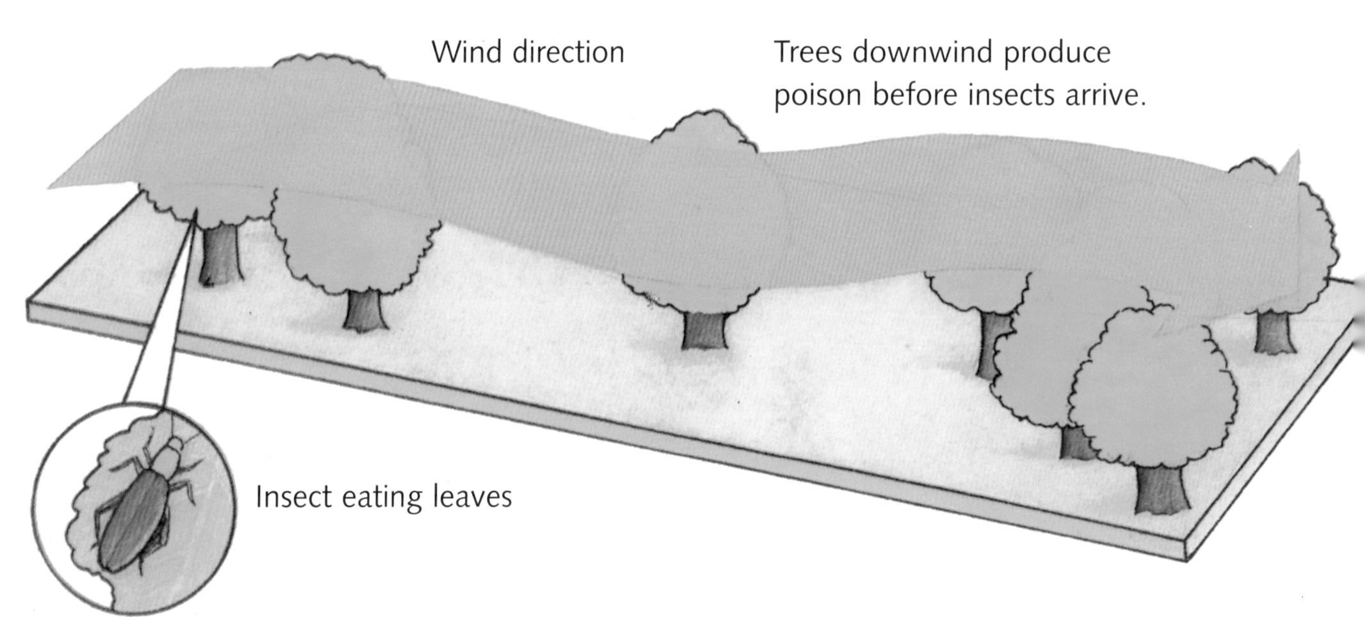

We usually associate communication with animals, but some plants can communicate. For example, some trees produce poison in their leaves when they are attacked by insects. Some of these trees can also warn nearby trees by passing a chemical signal through the air, enabling trees downwind to prepare for attack. In a similar way, plants also communicate with insects. If corn, cotton, or tobacco plants are attacked by caterpillars they emit chemicals that simultaneously attract parasitic wasps. These wasps eat the caterpillars and discourage other worms and moths from laying their eggs on the plants.

Monkeys use a number of vocal, visual, and tactile forms of communication.

Dolphins communicate using a system of whistles, squeaks, moans, and clicks.

Elephants show aggression by locking tusks and wrestling with their trunks.

LIVING WITH EACH OTHER

A living thing cannot live alone. It depends on other living things to supply the materials it needs to survive. We have seen how some animals rely on plants for their food. Animals that only eat plants are called "herbivores." Animals that eat other animals are called "carnivores." A series of living things that feed on each other makes up a "food chain." If one of the members of the chain is removed, all the others may be affected.

Often, several food chains interlink as many animals feed on a variety of plants or animals. The chains together make a "food web."

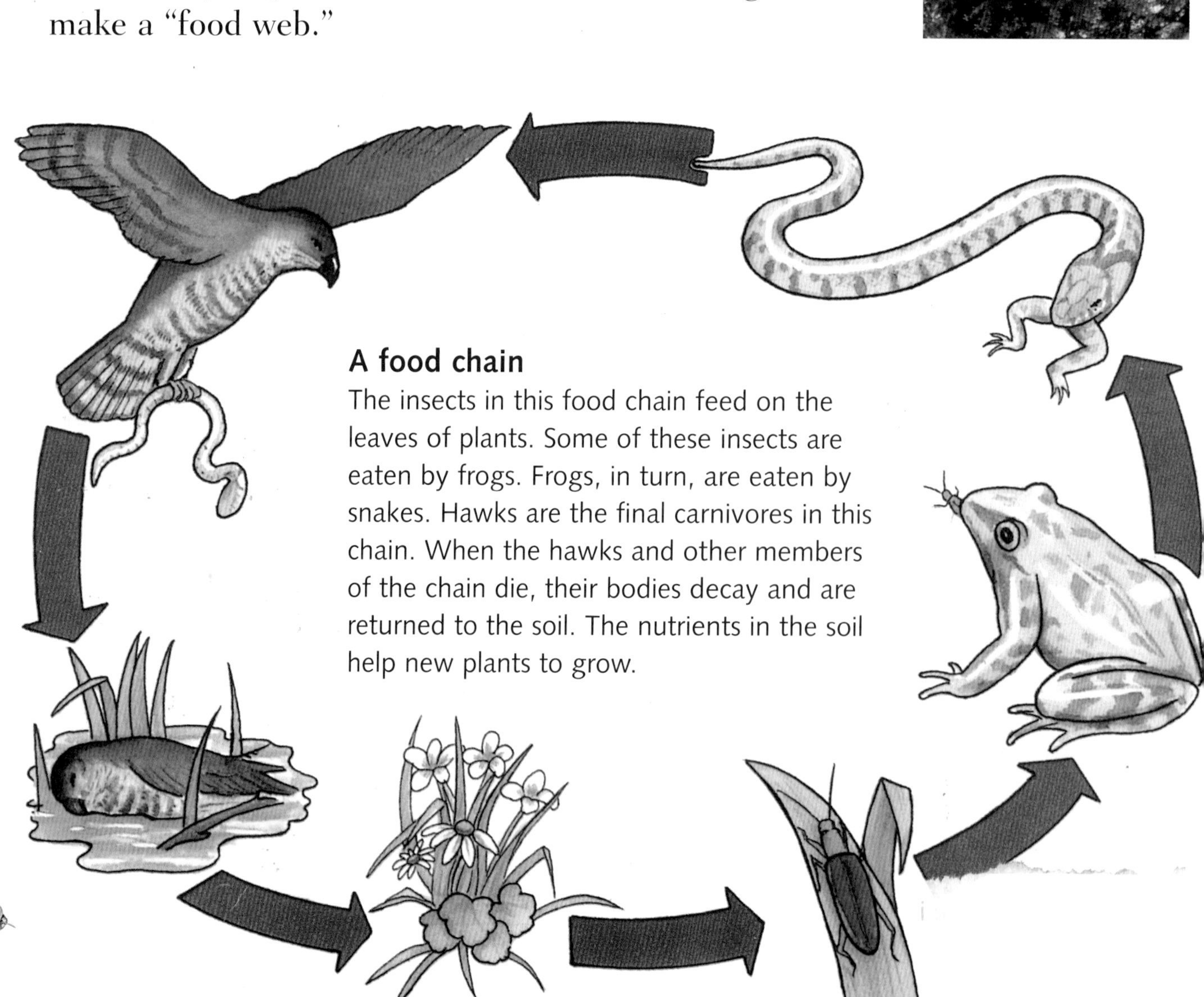

A food chain

The insects in this food chain feed on the leaves of plants. Some of these insects are eaten by frogs. Frogs, in turn, are eaten by snakes. Hawks are the final carnivores in this chain. When the hawks and other members of the chain die, their bodies decay and are returned to the soil. The nutrients in the soil help new plants to grow.

Kingfishers eat fish, but they themselves are eaten by larger birds and small mammals.

mpala herbivores are in danger of being caught by the lioness, a carnivore.

MAKE A "FOOD WEB" GAME

You can have a lot of fun making and playing this "food web" game with your friends. You will also learn more about living things and the way in which they depend on each other for food. All the living things in this game live in Africa. To make the game you will need some cardboard, scissors, tracing paper, pencils, crayons, and a ruler.

Rules

The game can be played by up to six players. There are 36 cards in the deck. All the players should be dealt an equal number of cards. If there are any cards left over leave them out of the game—but check that MAN is not one of these—he MUST be included in the game. The aim of the game is to get rid of your cards as quickly as possible. The winner is the first person to use all of his/her cards.

How to play

All players look at their cards. The person who has MAN starts by placing this card down, face upward. Players then take turns in a clockwise direction. Each player must, if possible, put down a card showing a living thing that EITHER eats OR is eaten by the living thing on the previous card. If you cannot go you miss a turn. In this game, MAN eats everything, so the second player will always be able to put down a card.

Making the cards

Start by tracing the MAN card. Then look at the diagram below and work out what information to put on the remaining cards. For example, you need to make three WARTHOG cards (shown by the number in parentheses). Follow the arrows and you will see that the WARTHOG only eats PLANTS and is eaten by LIONS. Don't forget, in this game, MAN also eats WARTHOGS!

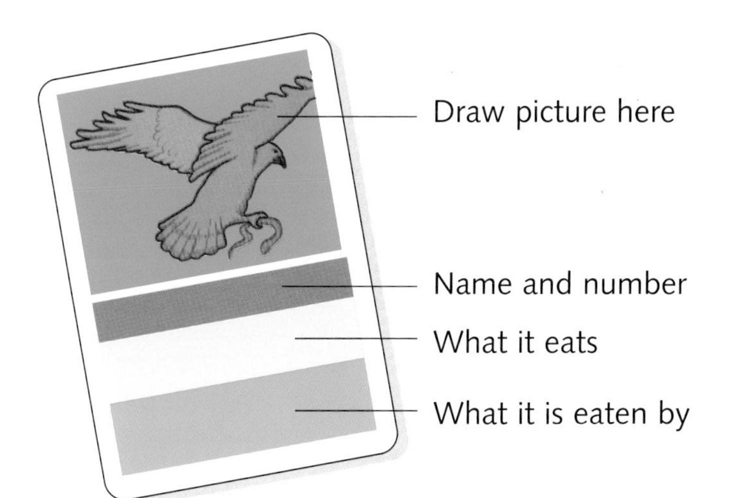

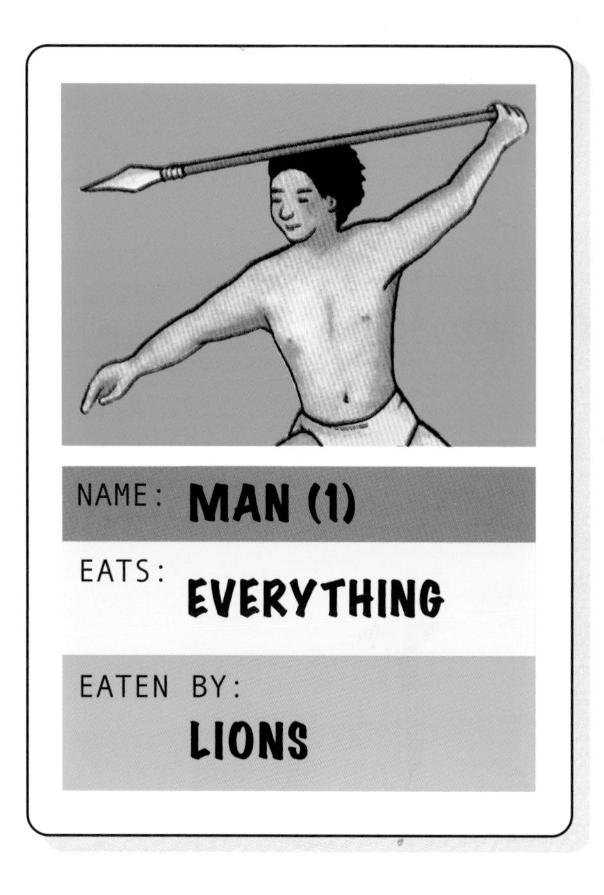

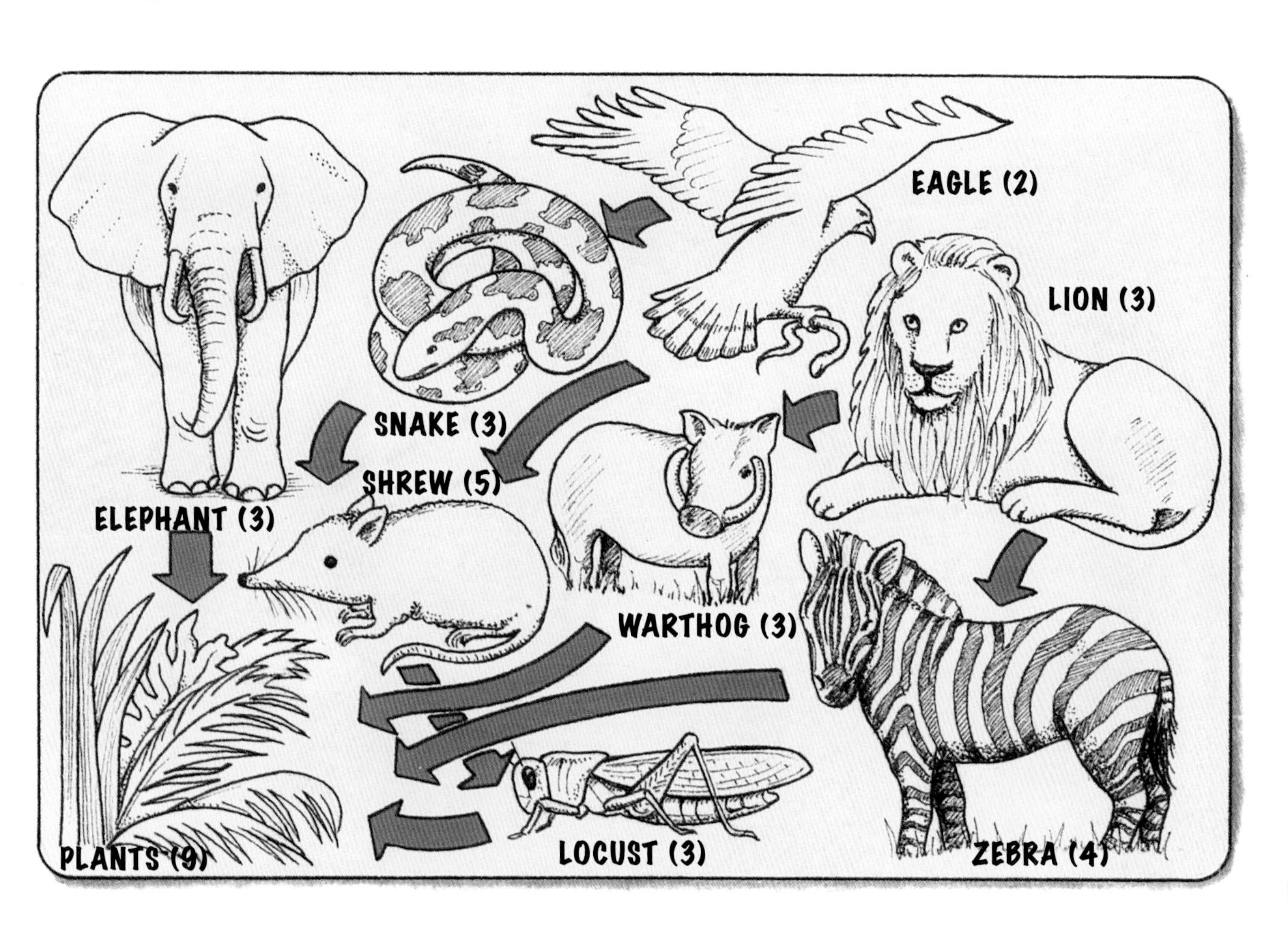

MORE ABOUT REPRODUCTION

Genes and chromosomes

Inside the nucleus of a human cell are 23 pairs of chromosomes. Along these chromosomes are "genes." Genes control everything about you, including the way you look. Each of your cells has the same genes. You inherit two copies of every gene—one from your mother and one from your father. That is why you look like your parents.

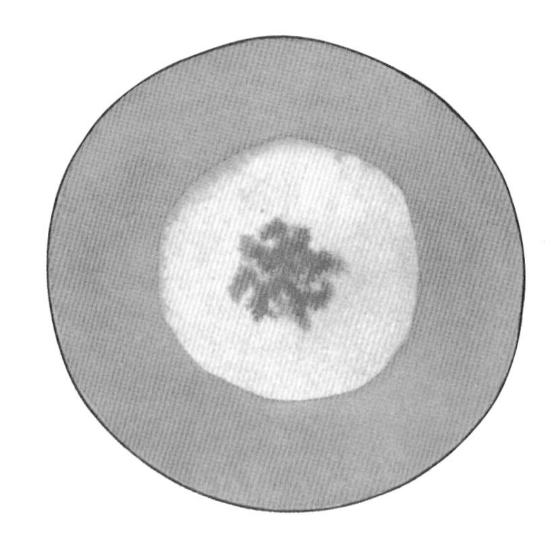

Inheriting eye color

The gene for brown eyes is known as a "dominant" gene—you only need to inherit one brown eye color gene for your eyes to be brown. The gene for blue eyes is weaker. A child will only have blue eyes if both inherited eye color genes are blue. This man carries a gene for brown eyes and a gene for blue eyes, while the woman carries only a gene for blue eyes. One of their daughters has blue eyes because she has inherited two blue eye color genes. The other daughter has brown eyes because she has inherited at least one brown eye color gene.

GLOSSARY

Camouflage
A colored or patterned appearance that disguises an animal within its surroundings.

Carbon dioxide
A colorless gas found in the air. Plants use it to make sugars and starches.

Cell
A very small part of living matter. Different kinds of cells do different jobs in the body of a plant or animal.

Dominant gene
A gene carrying information that is always expressed in a living thing, even if only one of its type is present in a cell.

Evaporate
To change a liquid into a gas. A liquid needs energy to evaporate and so absorbs heat from its surroundings.

Evolution
The way in which plants and animals have changed gradually over millions of years from simple to more complex forms.

Feces
The solid waste matter, derived from digested food and expelled from the body of some animals.

Hibernation
The winter sleep when an animal's temperature lowers and heartbeat slows.

Invertebrate
An animal that does not have a backbone.

Mammals
Animals that produce milk to feed their young.

Minerals
Substances in soil, water (for plants), or in food (for animals) that are needed for growth.

Nutrient
A substance that provides nourishment, such as minerals.

Oxygen
A colorless gas that makes up about a fifth of the air. Oxygen is essential to the lives of plants and animals.

Photosynthesis
The process by which plants produce oxygen and make sugars and starches from carbon dioxide and water, using energy from sunlight.

Pollination
The transfer of pollen grains from the male to the female parts of a flowering plant, for reproduction.

Sex cell
A cell that has to join with another cell before it will develop into a new plant or animal.

Skeleton
The hard parts of an animal that supports its body. A skeleton may be on the outside of the body (such as the shell of a crab) or on the inside (such as the bones of a human or a cat).

Species
A group of plants or animals that are like each other and that can reproduce together. For example, all dogs belong to the same species.

Urine
The watery liquid that carries waste substances out of the bodies of some animals.

Vertebrate
Any animal that has a skeleton of bone or cartilage, with a backbone.

Zygote
A cell formed in plants and animals by the joining together of a male and female sex cell. In the right conditions, a zygote will develop and grow into a new plant or animal.

INDEX

Photocredits Abbreviations: l-left, r-right, b-bottom, t-top, c-center, m-middle. All pictures supplied by Digital Stock except for: 1 — Corbis Royalty Free. 4tl, 6tl, 8tl, 10tl, 12tl, 14tl, 14tr, 16tl, 18tl, 18ml, 18bl, 20tl, 22tl, 24tl, 26tl, 28tl, 30t, 31t, 32t — Stockbyte. 4b, 8mr, 25tl, 25b — Corbis. 6tr, 10tr, 21, 23b — Photodisc. 6b, 7, 17b, 25tr, 28tr — Corel. 11 — Jack Dykinga/USDA. 12tr — Comstock. 13b — Picturepoint. 15tr — Bruce Fritz/USDA. 22tr — Keith Weller/USDA. 23t — Scott Bauer/USDA. 30b — Digital Vision.